baby MOOSE

KIM THOMPSON

CREATIVE EDUCATION • CREATIVE PAPERBACKS

CONT

ENTS

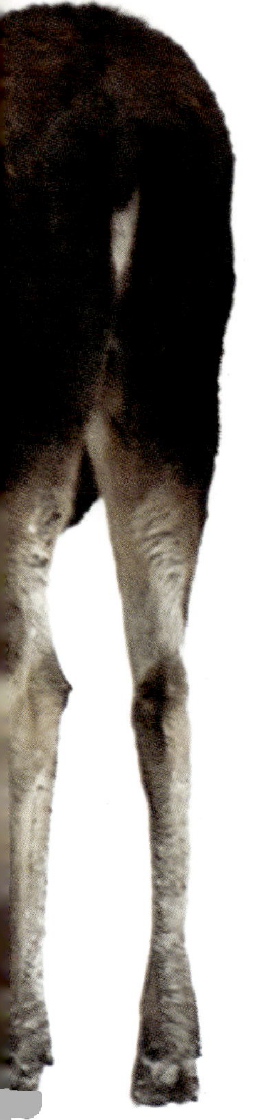

I AM A CALF.

I am a baby moose.

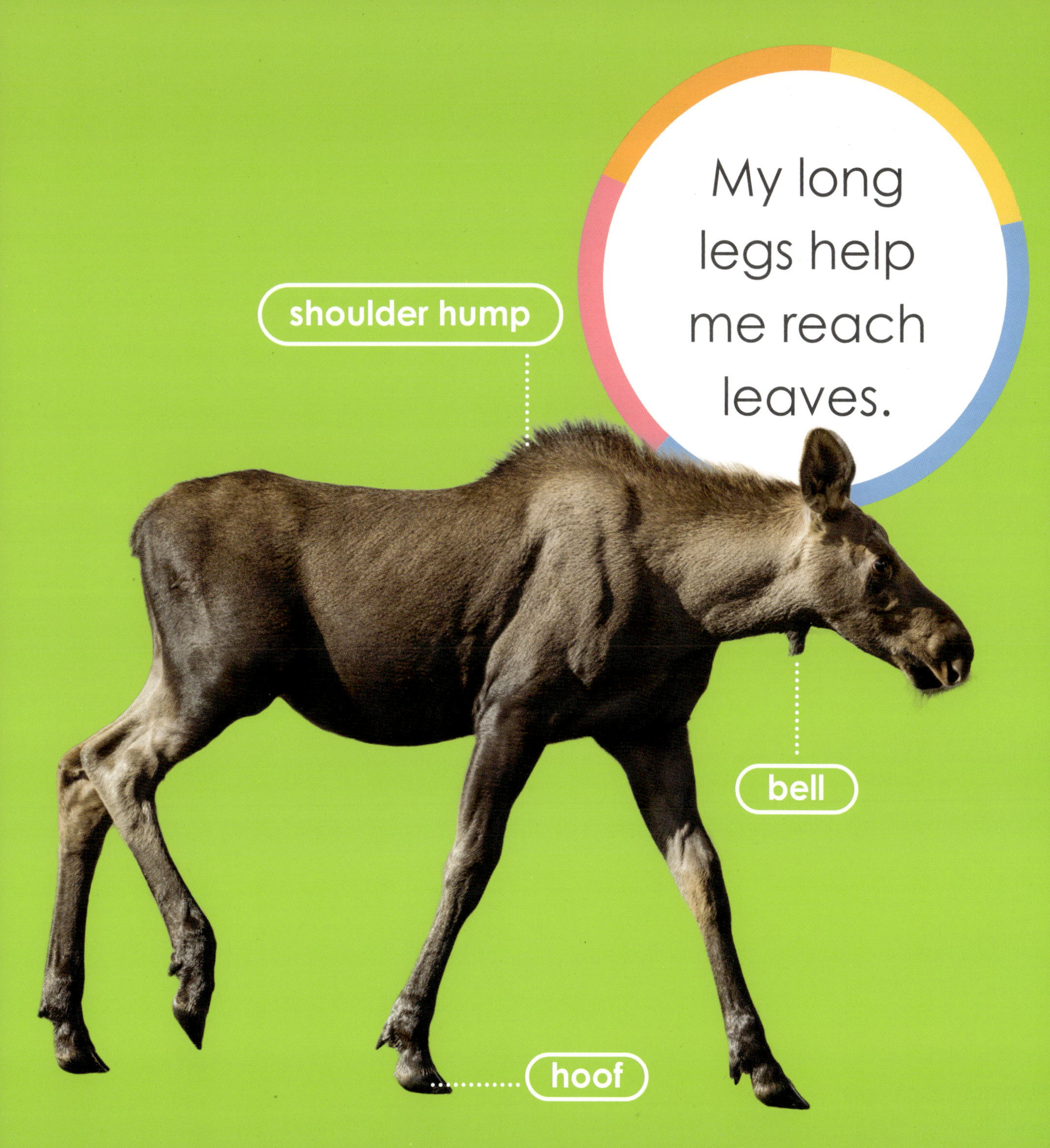
My long legs help me reach leaves.
shoulder hump
bell
hoof

My mom had two babies at the same time. We drink her milk.

I eat and grow all spring and summer. I will be big and strong when cold weather comes.

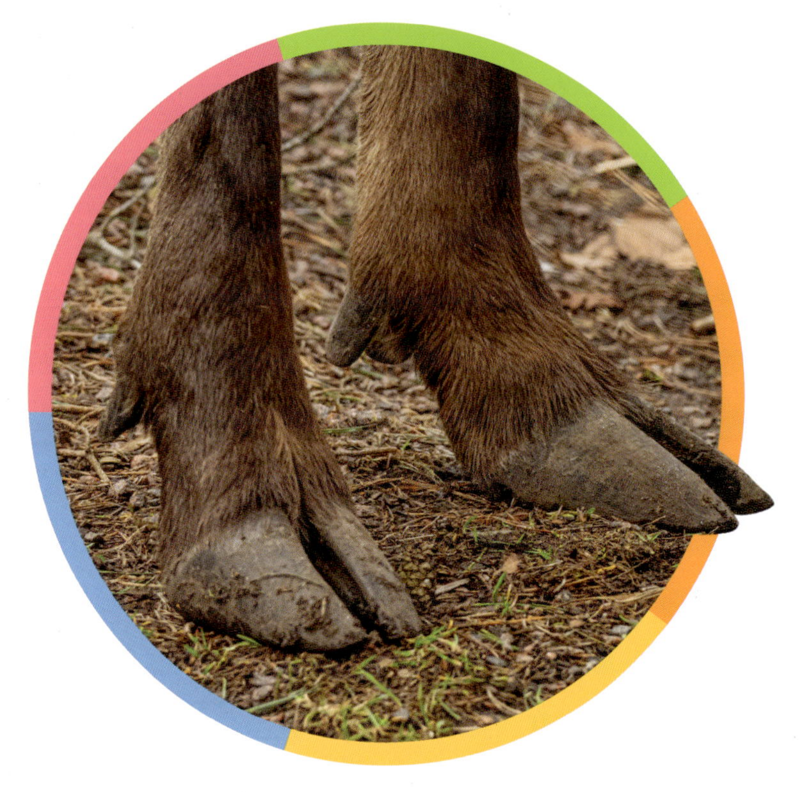

I **explore the** woods. I find twigs to eat. My hooves help me stand on mud.

I am a fast runner!

I am a boy moose, or a bull. I am one year old. My antlers are starting to grow!

One day, I will weigh as much as five refrigerators.

SPEAK AND LISTEN

Can you speak like a calf?

Baby moose mewl and whine.

Listen to these sounds:

https://www.youtube.com/watch?v=BlVPlgRARcg

MOOSE WORDS

antlers: branching, bony body parts on the head of a male moose

bull: an adult male moose

hooves: hard, nail-like coverings on a moose's feet

twigs: small, thin tree branches

READING CORNER

Castaldo, Nancy F. *The Wolves and Moose of Isle Royale*. New York: Clarion Books, 2022.

Gamble, Adam, and Mark Jasper. *Good Night Moose*. New York: Penguin Random House, 2020.

Riggs, Kate. *Moose (Amazing Animals)*. Mankato, Minn.: Creative Paperbacks, 2021.

INDEX

PUBLISHED BY CREATIVE EDUCATION AND CREATIVE PAPERBACKS
P.O. Box 227, Mankato, Minnesota 56002
Creative Education and Creative Paperbacks are imprints of The Creative Company
www.thecreativecompany.us

LIBRARY OF CONGRESS CATALOGING-IN-PUBLICATION DATA
Names: Thompson, Kim, 1970- author.
Title: Baby moose / Kim Thompson.
Description: Mankato, Minnesota : Creative Education and Creative
Paperbacks, [2026] | Series: Starting out | Includes bibliographical references and index. | Audience: Ages 4-7 | Audience: Grades K-1 |
Summary: "Introduce beginning readers to the world of baby moose with this life science starter. Includes photos, a labeled animal diagram, "Make a Noise" section, glossary, and further resources"-- Provided by publisher.
Identifiers: LCCN 2024043241 (print) | LCCN 2024043242 (ebook) | ISBN 9798889897507 (library binding) | ISBN 9781682778364 (paperback) | ISBN 9798889897637 (ebook)
Subjects: LCSH: Moose--Infancy--Juvenile literature.
Classification: LCC QL737.U55 T487 2026 (print) | LCC QL737.U55 (ebook) | DDC 599.65/71392--dc23/eng/20241129
LC record available at https://lccn.loc.gov/2024043241
LC ebook record available at https://lccn.loc.gov/2024043242

DESIGN AND PRODUCTION
Design by Rhea Magaro
Production by Beeline Media and Design, Inc.
Art direction by Tom Morgan

PHOTOGRAPHS by Alamy Stock Photo/Naturfoto-Online, cover; Dreamstime/Anagram1, 11, Ingemar Magnusson, 7, Jeanninebryan, 4, 6-7; Getty Images/Freder, 10-11, LudiLen, 2-3, Patrick J. Endres, 5; Shutterstock/Bohbeh, 13, Jennkuhn, 12, Jon C. Beverly, 14, RisingTimber, 9, Trygve Finkelsen, 8

Printed in India